The New Forest

its character and heritage

The New Forest

In 1079 King William I declared a vast tract of woodland and heath to the south of his castle at Winchester a royal hunting ground, designating it his New Forest. The term 'forest' referred not to trees, as in the modern sense, but to an area that was subject to Forest Law, reserving the pursuit of beasts within it exclusively for the king and his officers.

The poor soil of the area was not suitable for extensive agricultural use and, as a result, the 'forest' was sparsely populated. Nevertheless, the subjection of the area as such considerably curtailed the activities of the Saxon people who lived there. Dire penalties of death or mutilation were incurred by anyone caught in the act of even disturbing the king's deer, let alone obtaining venison or game. No smallholder could enclose land in such a way as to impede the free running of the royal quarry, and the need to preserve woodland cover meant restrictions on the cutting of timber for domestic use. However, certain customary common rights were allowed, including the grazing of livestock.

Under Magna Carta (1215) and Carta de Foresta (1217) the draconian Norman Forest Law was changed. In succeeding centuries the administration of the New Forest reflected varying conflicts and concerns of Crown and people. Yet, many of the rights of common still remain at the heart of today's New Forest. Ancient traditions coexist with modern forestry practice within a setting that is home to an impressive variety of natural habitats for wildlife, ensuring a forest that all may enjoy.

The boundaries of the New Forest have tended to waver throughout its history. Some of the settlements on the edge of the wooded areas and heathlands have been variously deemed to be in or outside its perambulation. In 1964 cattle grids were put in place to prevent ponies and cattle wandering out of the forest and the main roads were fenced off to protect animals from accident and injury. In 2004 the forest was designated a National Park, with a further defining of its borders.

The true outlines of the New Forest, however, are found in the unique character of its heritage and traditions, its scenery, its creatures and its ability to remain a natural haven in a changing world.

Early forest dwellers

Within the area there are many round barrows (burial mounds) that can be dated to the Bronze Age. There are also Iron Age earthworks dotted about the forest.

Roman pottery has been unearthed and some vestiges of Roman roads.

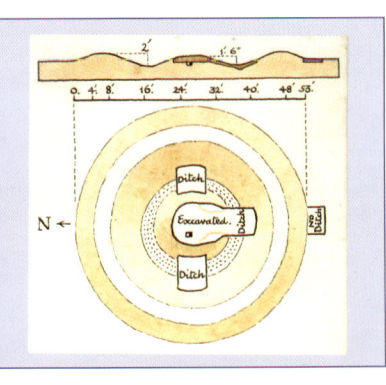

RIGHT: *Disc barrow on Ibsley Common, illustrated by Heywood Sumner.*

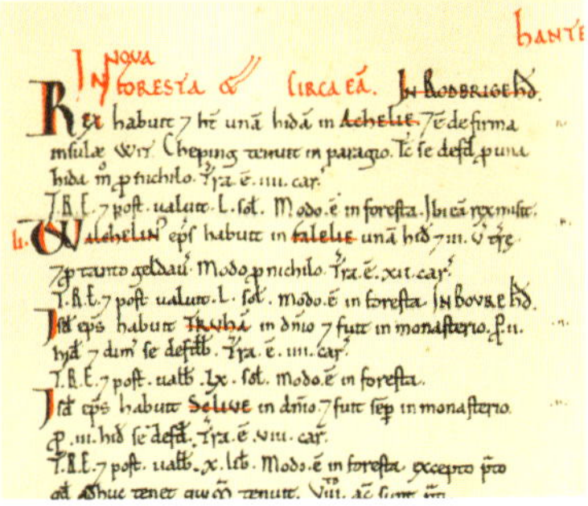

LEFT: *An Iron Age hillfort at Burley afforded defensive views over the surrounding countryside.*

ABOVE: *An extract from Domesday. The New Forest had a separate section at the end of the account of Hampshire.*

LEFT: *Iron Age bronze coins found at Iron's Hill, Lyndhurst, may be seen at the New Forest Museum.*

BELOW: *The death of King William Rufus is portrayed in a modern embroidery at the New Forest Museum, Lyndhurst.*

The Rufus Stone

King William II (known as Rufus because of his ruddy complexion) was killed by an arrow while hunting in the New Forest on 2 August 1100. There has been much debate as to who shot the arrow and whether the king's death was an accident. Chroniclers of the time do not suggest anything untoward or name a culprit, but tradition places blame at the hands of Sir Walter Tyrrell who fled to France soon after.

This historic event is commemorated by the Rufus Stone, a small monument, at Canterton, near Stoney Cross. The original stone was placed here in 1745. The inscription reads:

Here stood the Oak Tree, on which an arrow shot by Sir Walter Tyrrell at a stag, glanced and struck King William the second, surnamed Rufus, on the breast, of which he instantly died, on the second day of August, anno 1100.
King William the second, surnamed Rufus being slain, as before related, was laid in a cart belonging to one Purkis, and drawn from hence, to Winchester, and buried in the Cathedral Church of that City.

However, recent evidence supports the theory that the king may have met his end at Througham, south of Beaulieu.

The New Forest: Spring and Summer

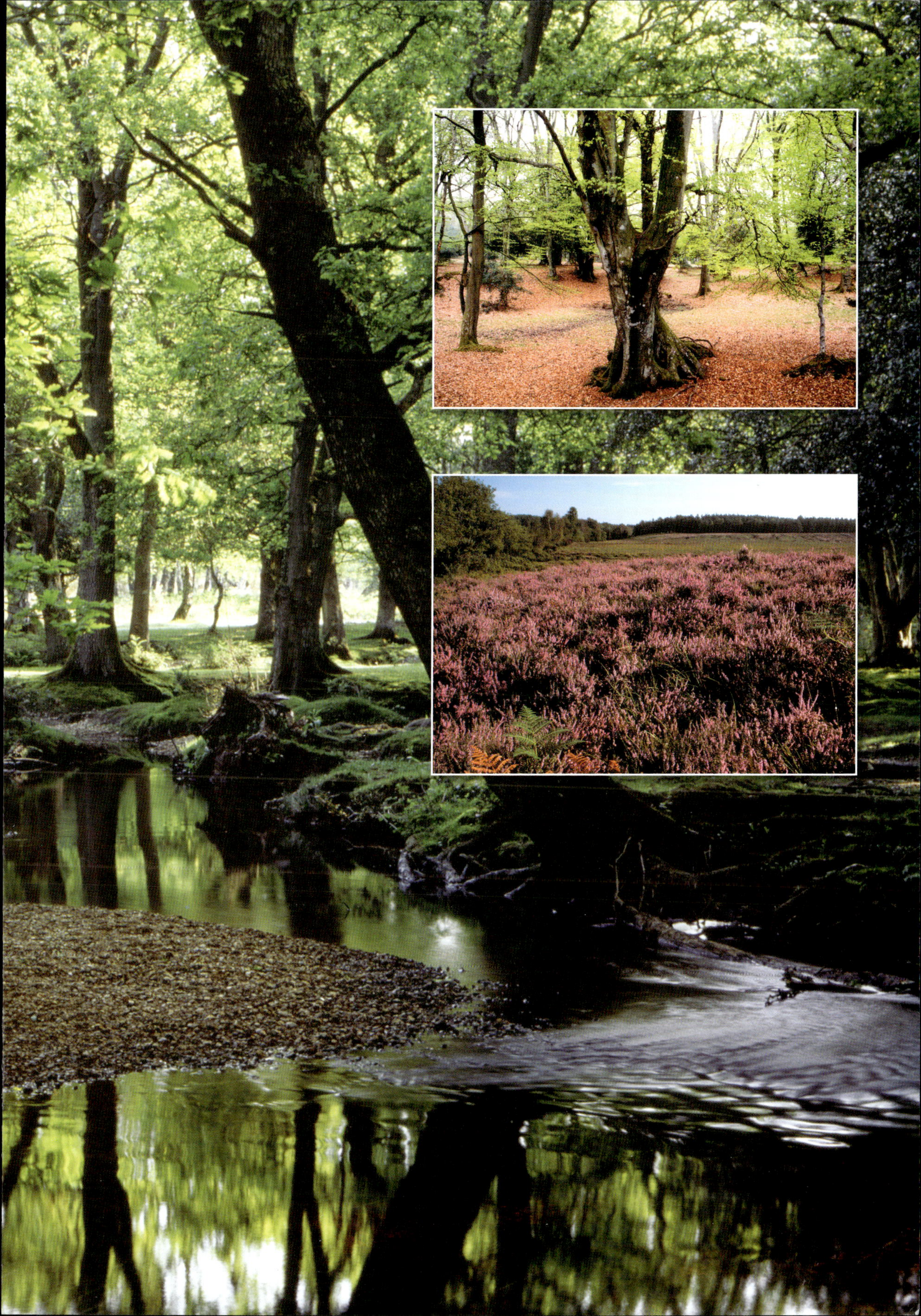

The New Forest: Autumn and Winter

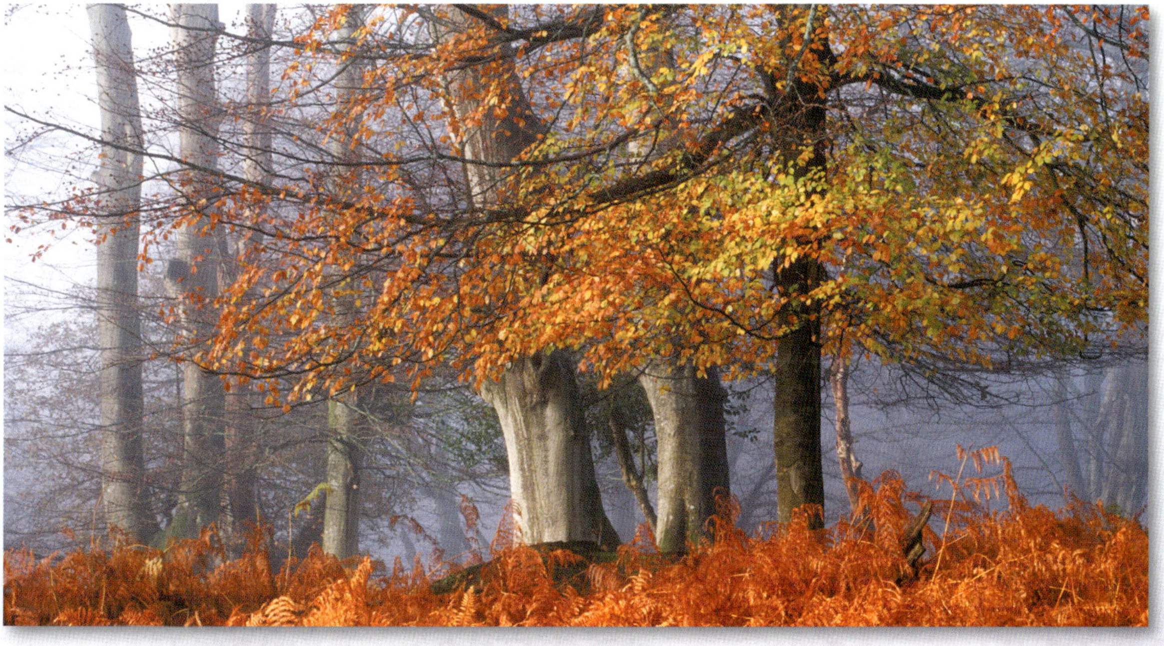

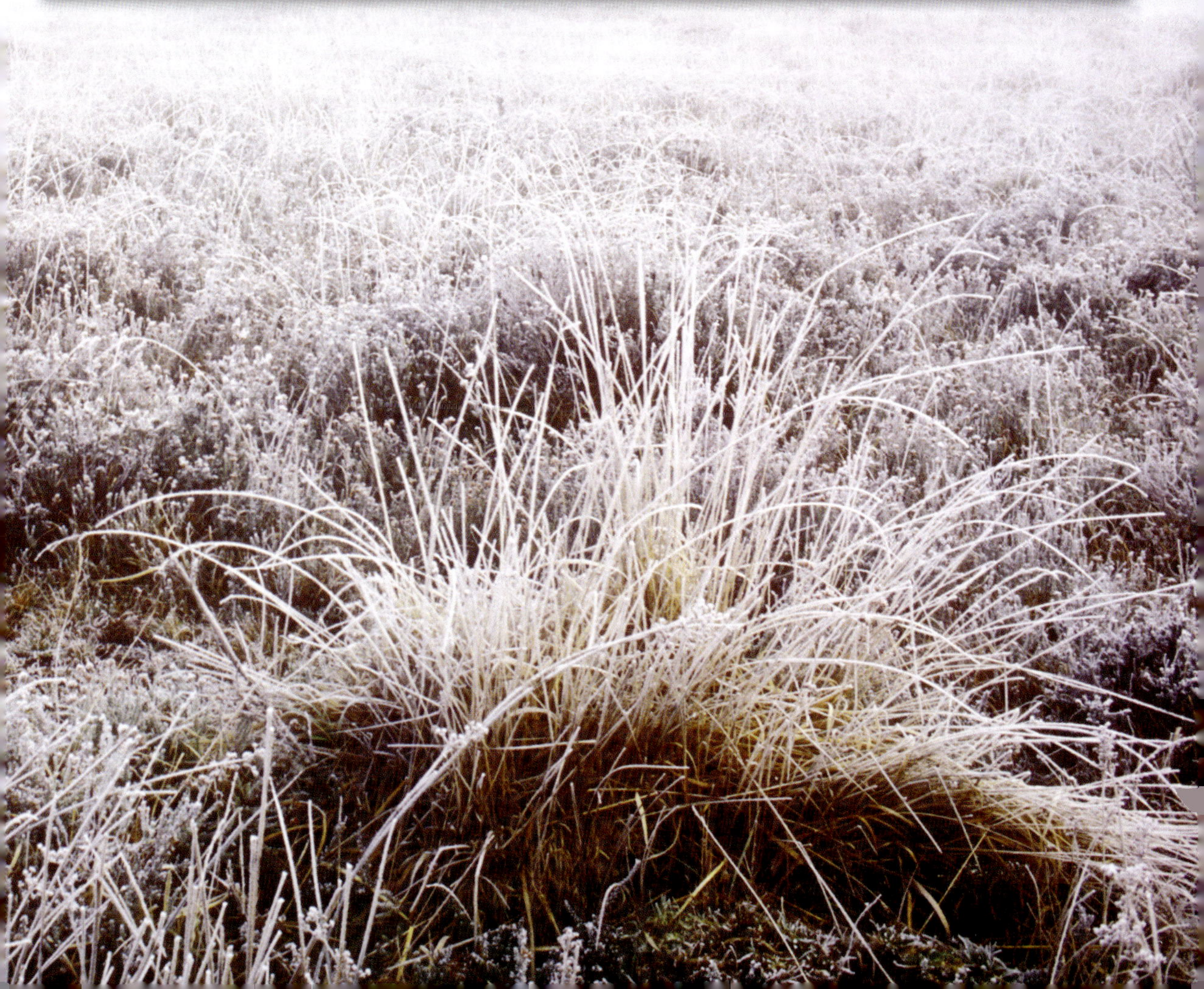

Commoners

It is 'commoning' that has essentially conserved the New Forest as a unique area. Commoners still depasture their livestock on the forest, and the grazing of these animals keeps the growth of scrub in check. Sadly, however, the numbers who follow this traditional way of life have declined in the face of modern pressures. Few have ever relied solely on commoning as a means of making a living.

A commoner is defined as someone who owns or occupies land to which rights of common are assigned. The rights are attached to the holding rather than to its owner and, in times past, the right of estovers was assigned to the hearth of the property. About 400 commoners still exercise their rights, which vary slightly from one holding to another according to the particular area. Many rights are held by properties that are outside the actual 'perambulation' of the forest.

The six common rights

Common of Pasture, the most important right, which allows the holder to depasture commonable animals on the open forest. Commoners pay a small fee for all cattle and ponies 'turned out'. In some areas commoners also depasture donkeys.

Common of Mast, the right to turn out pigs on the forest during what is called the 'pannage' season in autumn, when the acorns and beechmast have fallen from the trees. If ponies and cattle eat too many acorns they are liable to protein poisoning, which can prove fatal, but these seasonal fruits make good feed for pigs. Pannage must last for a period of not less than 60 days and dates are announced each year according to seasonal variations. In some areas of the forest pigs are allowed out throughout the year.

Common of Turbary, no longer exercised, the right to cut turf for fuel. The basic rule is that for every turf cut two adjoining ones are left, ensuring that an area is not depleted and the grass can regenerate.

Common of Fuelwood, sometimes called Estovers, the right to use wood to burn for domestic use. This right is attached to only a few properties. The Forestry Commission cuts the wood, stacking it in bundles known as 'cords', for the commoner to collect.

Common of Marl, the right to dig marl from registered forest pits. Marl is a lime clay that may be used as a fertiliser for poor soil. This right is no longer exercised.

Common of Pasture of Sheep, a right granted to only a few commoners to depasture these animals. The distribution relates to the holdings of religious houses in medieval times.

LEFT: *Commoners gathering before a 'drift' to round up ponies for branding, tail marking and veterinary checks.*

ABOVE: *Pigs are turned out for pannage during the autumn.*

TOP RIGHT: *Depastured cattle, like ponies, have free range of the unenclosed forest.*

RIGHT: *Commoners make good use of the lowland heath as ideal sites for beehives.*

FAR RIGHT: *A typical commoner's smallholding to which rights of common were assigned.*

BOTTOM RIGHT: *Wood cordage, cut and stacked by the Forestry Commission.*

Traditional customs

Commoners have always made the most of the natural resources allowed to them. In addition to rights of common, anyone who lives within the forest in a property that was built before 1850 may gather fallen branches for firewood, as long as they do not use a vehicle to transport them.

When bracken dies back in the autumn it makes good bedding for ponies and cutting it allows the grass to grow, which is essential for grazing. In hard winters holly cuttings provide additional feed.

The gorse and heather provides nectar for bees, and beehives are placed in the forest in the summer in certain places for a fee payable to the Forestry Commission.

A commoner's home

The traditional building material in the New Forest was cob, a mixture of clay and straw. A typical small holding would have had an outside privy and outbuildings such as a pigsty and sheds for chickens, geese and larger livestock.

Commoners were largely self-sufficient, growing their own fruit and vegetables and collecting honey from their own bees. A small paddock adjoined the land so that ponies and cattle could be brought in off the forest.

Verderers and Agisters

The officials charged with enforcing Norman law were known as verderers, the name coming from their responsibility for the 'vert', meaning 'greenery'. Their job was to protect the forest, ensuring a beneficial habitat for the deer and a suitable environment in which to hunt them. Infringements were investigated by the verderers at a Court of Attachment and evidence against an individual could be referred to a higher court and appropriate sentencing. Closely allied to this body was the Court of Swainmote, which enabled the verderers to administer the seasonal pannage of pigs and the removal of cattle from the forest during the 'heyning' months of winter when all grazing was reserved for the king's deer and for a 'fence' month in the summer when the does gave birth to fawns.

Since the late nineteenth century these two courts have been combined and the verderers today are concerned with the protective management of the wider interests of the forest and its commoners. They control the development of the environment and have overall jurisdiction over the welfare of animals depastured on the forest.

There are ten verderers – five are elected and five are appointed as representatives of the Crown, the Forestry Commission, the Countryside Agency, Hampshire County Council and Defra. Elected verderers must occupy at least one acre of land to which common rights are attached.

The verderers sit in open court at regular intervals throughout the year and any member of the public may make a 'presentment' to them, raising any issue concerning the forest. These sessions are held in the Verderers' Hall at The Queen's House in Lyndhurst.

The day-to-day practical management of the forest with regard

LEFT: *Royal coat of arms of Charles II who was responsible for the rebuilding of the then King's House to which the Verderers' Hall is attached.*

BOTTOM LEFT: *The 'stirrup of Rufus' was used to determine whether a dog would be a threat to royal hunting. Any dog that was too large to pass through the stirrup was maimed to prevent it chasing deer.*

BELOW: *The agisters of the forest wear a distinctive dark green livery on formal occasions.*

The New Forest | 11

to animals is carried out by five agisters employed by the verderers. It is the Head Agister who opens each meeting of the Verderers' Court by raising his right hand and declaiming:

'Oyez, oyez, oyez!
All manner of persons who have any presentment or matter or thing to do at this Court of Verderers let him come forward and he shall be heard.
God Save The Queen!'

TOP: *The Court of Verderers in session.*

ABOVE: *The court is held in the Tudor Verderers' Hall, which adjoins The Queen's House.*

Agisters at work

The agisters work full time on the forest and keep a careful check on the welfare of all livestock within their particular area. For each animal that is depastured on the forest the owner must pay a small annual fee towards administration, known as a 'marking fee', and this is collected by the agisters. In the case of ponies, proof of payment is denoted by the agister clipping the animal's tail to a set pattern according to area, while cattle are tagged. Commoners are also required to brand depastured animals, the individual owners' marks being registered with the verderers.

A pony's tail is cut to show the area in which the animal is depastured.

Ponies

Ponies are part of the scenery of the New Forest. Over the centuries commoners have used ponies for riding and for carriage and cart driving, but the ponies depastured on the open forest are free-running, unridden animals. Since the forest roads pass through common land, ponies have right of way.

As herd animals, ponies like company and tend to cluster in small family groups. Mares give birth to foals on the forest and ponies usually prefer to stay within a wide familiar area unless moved some distance away.

Other breeds have been introduced to improve the native stock at various times, but since the establishment of the New Forest Pony and Cattle Breeding Society in 1938 no outside blood has been allowed in registered ponies. This is effectively controlled by the verderers, who oversee the running of registered stallions on the forest. To keep the forest-bred stock healthy, stallions are moved every three or four years.

The welfare of animals is a prime concern of the agisters. In winter there may be a scarcity of grazing and by spring the ponies' condition can quickly deteriorate if not carefully observed. Agisters may ask a commoner to remove any animal in poor condition from the forest.

Pony 'drifts' are held each year in late summer and autumn, when both commoners and agisters ride out across the open forest to round up the animals. Herding the animals together in a pound allows several jobs to be done at once with help at hand. Tails are clipped by the agister as a receipt of the annual marking fee and brands are applied to the ponies so that ownership can be determined. Foals are rounded up with their mothers and are assumed to be the property of the owner of the mare. Ponies may be given a close-up veterinary check and wormed. Any colts are taken off the forest before they can breed, as are ponies that are to be sold or kept on a holding through the winter.

Pony sales are held several times, mainly in the latter part of the year, at stockades at Beaulieu Road. New Forest ponies make good riding and driving ponies for they generally have placid temperaments, are already used

ABOVE: *Regular pony sales are held at Beaulieu Road and are popular with commoners and visitors.*

LEFT: *Young ponies on the forest.*

to the presence of traffic, and are fairly easy to train. They are hardy animals, accustomed to living with the elements throughout the year.

Visitors are asked not to feed the ponies since this attracts them towards the road. Remember, too, that these animals are not tame – the gentlest looking pony can also kick and bite!

> **CAUTION!**
> Sadly, many animals are killed or maimed by car drivers. Ponies and cattle are likely to cross in front of traffic at any time.
>
> There is a maximum speed limit of 40mph on unfenced forest roads, but accelerating to this speed is often inappropriate, especially at night.

TOP: *Ponies are branded with their owner's mark.*

LEFT: *Following the drift, ponies are held temporarily in a pound.*

RIGHT: *The pony round-ups are controlled by the agisters of the forest.*

BELOW: *Ponies are driven by riders towards the pound.*

Ancient and Ornamental Woodlands

The areas of native woodland that are largely unenclosed in the New Forest are known collectively as the Ancient and Ornamental woodlands. Continuous grazing by animals has produced characteristic leafy glades with little scrubland. These woods are dominated by mighty oaks and beeches, many of which are hundreds of years old, but there is also yew and a good variety of shorter-lived trees such as rowan and birch. On the fringes of these broadleaved woods are holly groves or 'holms', an important animal food resource in the forest.

Originally these woodlands were preserved as covert for deer. Some undergrowth was cut as feed, while 'fallen' wood was removed by commoners for domestic use. In order to protect growth some areas became coppices, temporarily legally enclosed with a ditch and hedge or fence. This kept out deer and livestock, and allowed the controlled cutting of fast-growing small trees and shrubs as well as enabling saplings to develop.

The forest also provided timber for the use of the Crown and fortifications for the defence of the realm, but by the beginning of the eighteenth century this resource had taken on increasing importance. Timber, particularly oak, was required by the Navy for the building of ships and parts of the forest were enclosed to protect this valuable material. Over the next two hundred years new trees were planted in an effort to regenerate the woodland and augment the depleting stocks.

ABOVE: *In autumn the humid conditions in the ancient woodlands allow a wealth of fungi to flourish.*

RIGHT: *The ancient and ornamental woodlands of the New Forest are the largest area of natural deciduous woodland in Britain.*

LEFT: *A leafy forest beech glade in summer.*

The pressure of supply became less acute in the middle of the nineteenth century when wooden ships were replaced by ironclad vessels. The oak plantations established for shipbuilding have now reached maturity, but for the most part are preserved for their amenity value and as an important haven for wildlife. Today much emphasis is placed on restoring these pasture woodlands of broadleaved trees.

The Knightwood Oak

The Knightwood Oak at Bolderwood has a girth of over 7m (23ft), and may be about 500 years old, possibly making it the oldest pollarded oak in the forest. The practice of pollarding, to remove branches to provide fodder for the deer, resulted in a spread of growth. It was prohibited at the end of the seventeenth century since it did not provide timber that was suitable for shipbuilding.

Charcoal burning

Charcoal burner's huts were once a feature of the forest. The ancient craft of charring small branches for charcoal dated back to the ancient Britons. As coal became more widely available for use in industry the market for charcoal all but disappeared, and so did this traditional way of life.

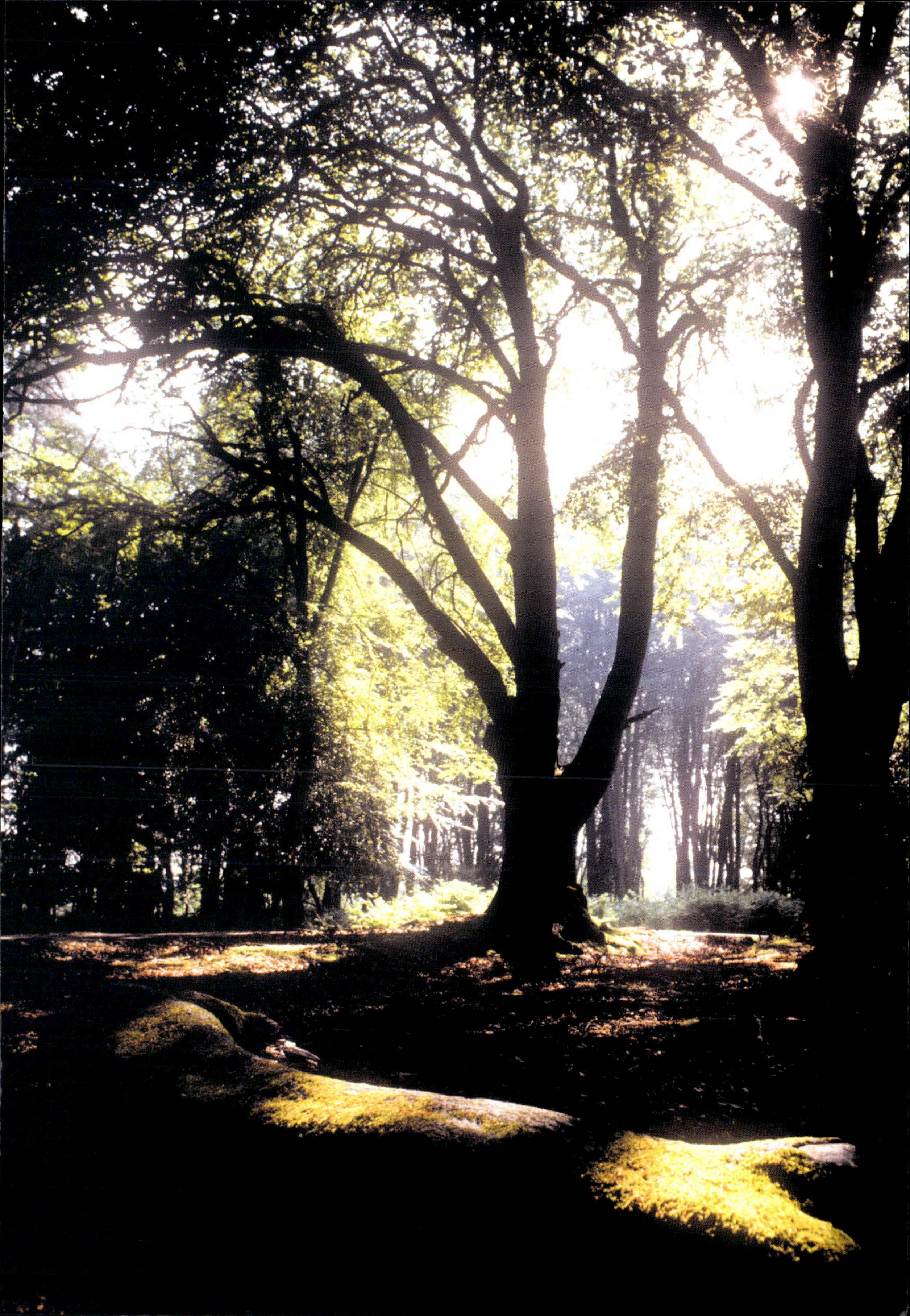

LEFT: *Badgers emerge from their setts under cover of darkness to forage for food.*

BELOW: *Woodland glade in spring.*

WOODLAND WILDLIFE

The woodlands of the New Forest provide a habitat for much wildlife. Foxes and badgers make their home here, though these creatures prefer to keep under cover if they are aware that they are being observed. Foxes are seen more frequently by the casual visitor.

Both foxes and badgers make their homes underground, but they have rather different housekeeping habits. The fox's earth may be dug where there is a convenient hollow, perhaps by enlarging a disused rabbit warren. The ground around the entrance is likely to be littered with the remains of previous meals such as rabbits, mice, voles, squirrels and birds.

Badgers prefer to make their sett on a suitable wooded bank, with a series of linked tunnels. These creatures are scrupulously clean, and the adult animals regularly bring in fresh bedding of bracken and leaves to replace used material. Badgers are omnivorous and the forest offers a varied diet of small rodents, snails, frogs and any earthworms they can find, plus berries and fruits.

There are many smaller mammals ranging from the ubiquitous grey squirrel to the less glimpsed weasel, which preys on rodents. The woodland floor provides plenty of food in the form of acorns, beechmast, berries and seeds for mice, voles and shrews whose populations rise and fall according to fluctuations in food supply.

These small rodents are themselves prey for owls, which nest in tree hollows and pierce the silence at night with their characteristic calls. The area is also good hunting ground for several species of British bats, which appear at dusk to search for flying insects.

Among woodland birds that thrive here are woodpeckers and jays. Jays, rather like squirrels, establish a hoard of food that sometimes goes unused, forming another link in the chain of regeneration on the forest floor.

As trees age, so branches fracture with crevices and holes that provide treetop homes for birds and many insects. On the ground moisture seeps into fallen branches and leaves, allowing fungal spores to germinate and set off the rotting process in the dead material. Decomposition attracts an abundance of insect life and microscopic organisms, which utilise nutrients released by the many fungi.

The damp woodland floor provides ideal growing conditions for ferns and mosses. A rich variety of lichens can also be found; these fungal plants are very slow to grow and the ancient trees of the forest offer an undisturbed location.

TOP: *The sun's rays pierce the oak canopy on a misty morning, supplying much needed light and warmth to flora and fauna.*

ABOVE: *The beautiful, but poisonous, fly agaric brings a flash of colour to the forest floor.*

RIGHT: *The ancient woodland provides both habitat and a source of food for the tawny owl.*

Deer

Deer are shy animals and tend to keep to the quieter parts of the forest during the day, but at dusk and dawn they venture near the roadsides. They are herd animals and move together in groups, rarely alone. The forest woodland provides good cover and shelter from the elements.

As ruminant animals, deer enjoy a diet of vegetation that includes huge quantities of foliage and grasses. They also tend to nibble at saplings and young trees, and it was for this reason that the Deer Removal Act was passed in 1851. Many animals were killed, but the keepers could not manage to destroy the whole population as they moved onto private lands. This, together with a new recognition of the role of deer in the pastured landscape, ensured their survival. There are four types of deer to be found in the New Forest today – fallow, roe, red and sika.

It is believed that fallow deer were first introduced into Britain by the Romans. In the seventeenth century the deer census in the New Forest reported a population of 7,500 fallow. Today there are about 1,700 of these animals.

Fallow deer are about 1m (3ft) high. In summer they have a light chestnut-coloured coat with white spots, but this changes to an unspotted grey-brown in winter. They have a distinctive black and white rump.

Roe deer, which are native to Britain, are rather smaller, measuring only 60cm (2ft) or so. Their winter coat is a grey-brown colour, but this also changes with the summer moult to a bright red-brown. The rump is white and the roe also has a white patch under the chin. A current count of roe deer in the forest gives a number of about 365.

At 1.2m (4ft) high, the red deer is the largest wild animal in Britain. Red deer have never formed a large population in the forest and

Deer Viewing at Bolderwood

There is a purpose-built observation platform at Bolderwood that allows the viewing of a herd of fallow deer. Animals are fed daily here between Easter and September by the local keeper, so they tend to remain in this area. They are not fed during the autumn rutting season or the winter culling season when the deer are likely to wander and find plenty of food elsewhere.

ABOVE: *Fallow deer are the most common deer in the forest and occasionally one of the animals may be white.*

RIGHT TOP: *A group of fallow does.*

RIGHT: *Roe deer doe and fawns. Roe deer are the smallest species of deer found in the New Forest.*

FAR RIGHT: *The age of a buck may be gauged by the shape and size of its antlers.*

LEFT: *A magnificent red deer, one of only a small number in the forest.*

descended from those given to the 2nd Baron Montagu at Beaulieu by King Edward VII at the beginning of the twentieth century. Today the sika in the forest number about 90.

Deer numbers are still carefully controlled by the keepers as these beautiful animals can cause much damage. In many wooded areas you may see evidence of their presence by an even browse line that indicates the height they can reach to eat.

Bucks grow new antlers just before the rutting season and the deer may rub against bark and saplings to remove the 'velvet' that covers the new antlers. Rutting for the roe deer takes place in late summer, and varies for the other species from early to very late autumn.

in the 1940s they had disappeared completely. However, a small private herd escaped from the grounds of Old House, Burley, in the 1960s and these have now increased to about 120.

Sika deer are of similar size to red deer. Their summer coat is a chestnut brown, spotted with white, but this becomes greyish in the winter and the spots disappear. Originally a Japanese species, the New Forest sika are

Antler development

Bucks grow their antlers anew each year. The shape develops as the animal gets older and the antlers are a good indication of the age of an animal.

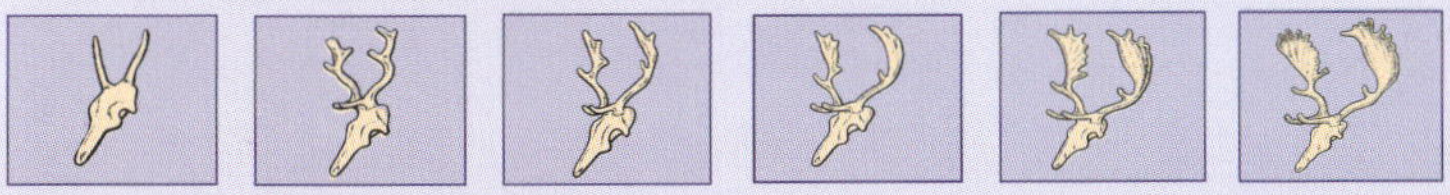

| Pricket (Yearling) | Sorel (2 year old) | Sore (3 year old) | Bare Buck (4 year old) | Buck (5 year old) | Great Buck (Over 5 years) |

The Inclosures

The first areas of woodland that were enclosed for timber regenerated themselves naturally, but at the end of the seventeenth century it was realised that stock needed to be replenished by planting. In 1698 an Act was passed that gave the Crown 'rolling' powers of enclosure. This meant that once trees in one area had grown to a size where they could not be damaged by grazing the fences were removed and other parts of the forest could be enclosed for new planting.

The 'rolling' powers proved to be of some contention between the authorities and the commoners, who quite reasonably viewed such activity as removal of their pasture land. Eventually, in the late nineteenth century, the special character of the New Forest was more generally recognised. Parliament approved legislation that halted the felling of ancient woodland and satisfied the rights of the commoners. Furthermore, the Verderers' Court was no longer required to serve the Crown, but was to uphold the commoners' interests and the traditional nature of the forest. In 1924 the Forestry Commission took over the practical management of the woodland. Today both parties work in cooperation, along with other interested bodies, with the aim of a balanced environment.

Conifer woodlands

The growing of trees, whether for timber or amenity value, is a long-term process. Hardwoods do not reach maturity for 120 years or so, while softwoods need to grow for at least 60 years.

In the eighteenth century Scots pine was planted in the New Forest for ornamental purposes and in the early nineteenth century as 'nurse' protection for hardwood trees. Many of these softwood trees were harvested while quite young for use as pit props. Later pure conifer plantations of pine, spruce and fir trees were established to provide timber for telegraph poles, fencing, pulp, logs for fuel, and for furniture.

The nature of these conifer enclosures is quite different from the broadleaved woodlands. Less light reaches the forest floor, which is covered with a carpet of fallen needles, and this habitat suits fewer types of plants and creatures. Nevertheless, the cones provide food for some species of bird, particularly the crossbill; the goldcrest nests here too. Plants such as foxgloves may sometimes grow at the edge of the track, where conditions are more likely to encourage vegetation. Wood ants make good use of the pine needles to make large cone-shaped ant hills.

ABOVE: *A wood ants' nest is a site of constant activity. The ants may burrow to a much greater depth than the pile of pine needles seen above ground.*

LEFT: *Felled timber is stacked along the edges of the inclosure rides for easy transportation.*

The foresters

In years gone by the regeneration of oak relied on the planting of acorns by hand, followed by other seeds, nuts and berries that would provide a protective shrub layer.

Today saplings are planted and protected from browsing wildlife by plastic tubes. Modern forestry practice and silviculture involves the use of efficient machinery and equipment for thinning, felling and sawing trees for timber.

The foresters' duties also involve conservational and recreational aspects of the forest and the maintenance of its infrastructure of bridges, paths, tracks, drainage, signs, car parks and camping sites.

The keepers

In the medieval forest the keepers were the guardians of the deer and enforced the severe laws that protected them. Over the centuries, however, their role changed as the preservation of woodland for timber became the priority. In the nineteenth century the keepers had to implement the removal of deer from the New Forest. This not only proved to be impossible, but it was clear that the deer were an essential link in the ecology of the open land used as pasture for livestock.

Today the keepers are employed by the Forestry Commission and are responsible for a stable, healthy deer population, with a census of animals being taken annually. They patrol their own beat and are concerned with wildlife, conservation and recreation facilities within that area as well as protection of the deer against poaching.

ABOVE: *A hobby feeds its chick in the safety of the upper branches of the conifers.*

BELOW: *The softwood plantations are principally a timber crop, rather than trees grown for amenity, but nevertheless they are home to much wildlife.*

The Lawns

LEFT: *For centuries commoners have depastured animals on the sweet grassland of the forest lawns.*

RIGHT: *The lawns are a haven for the naturalist, for they have a mix of habitats that range from grassland to stream, bordered by woodland, scrub and heath. The result is a rich diversity of flora and insects.*

Much of the forest consists of acidic grassland, bordering on heathland, but there are also expanses of neutral grassland with a wider range of grasses and other plants. These areas, or 'lawns', are usually in the flood plains of the streams that flow through the forest and they offer the best quality grazing. The name is apt, for the continuous grazing of ponies and cattle keeps the grass closely cropped.

While there may be woodland or heath to one side of a lawn, different habitats are supported alongside the streams and ponds, which are often edged with alder trees. The appearance and setting of the lawns make them a popular focus for recreation. A rich diversity of wild flowers grows along the banks of the streams, with yellow flag, purple loosestrife, water plantain and water crowfoot. Towards the woodland crab apple, honeysuckle and bramble grow. All these plants provide nectar for butterflies and other insects. There has been a decline in butterfly numbers in the New Forest, but many species may still be seen.

The streams and ponds support an abundance of wildlife. Generally the streams are clear-running with gravel bottoms against which shoals of minnows may be spotted. A territorial heron is usually a good indicator of occasional larger fish. Aquatic insects include water boatmen, water skaters and whirligig beetles, and the ponds are breeding grounds for many species of dragonfly and damselfly. Further upstream fallen branches and leaves cause dams and pools that slow down the rate of flow, helping to protect the stream banks from further erosion.

Amphibians are born in the water and return there to breed, but live mostly in damp places on land. The New Forest is home to both the common frog and the common toad, both of which are frequently found. A smaller species of toad, the natterjack, is more rare. Frog and toad spawn consists of thousands of eggs, but only a small percentage survive after the tadpole stage to maturity as adults for they have many predators from water, land and air. The three species of British newt are found here, too – the smooth newt, the palmate newt and the great-crested newt. These small amphibians have tails and are often mistaken for lizards. Like toads, the great-crested newt has a warty skin with glands that secrete an unpleasant-tasting substance.

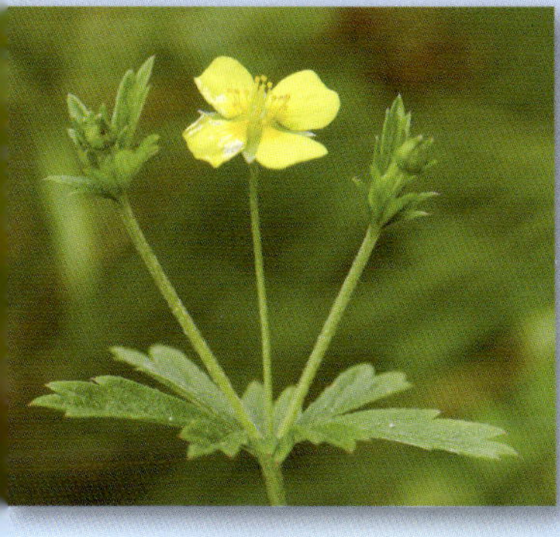
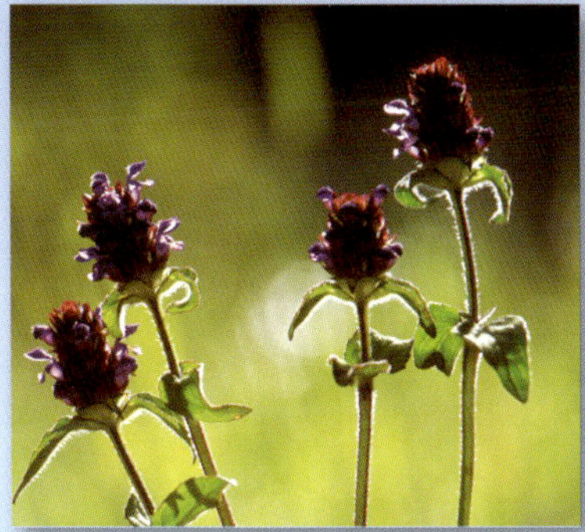
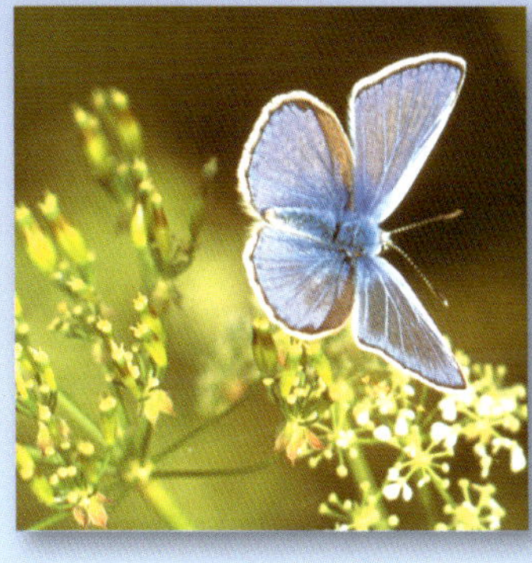
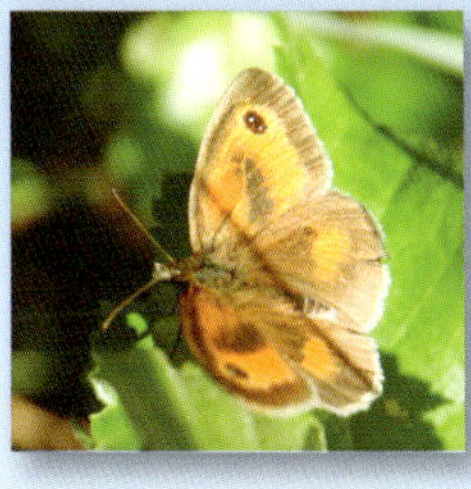
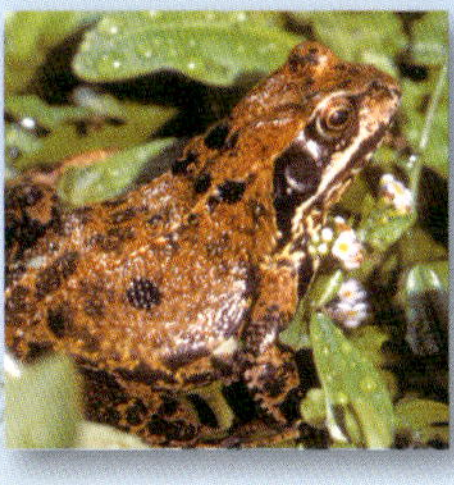
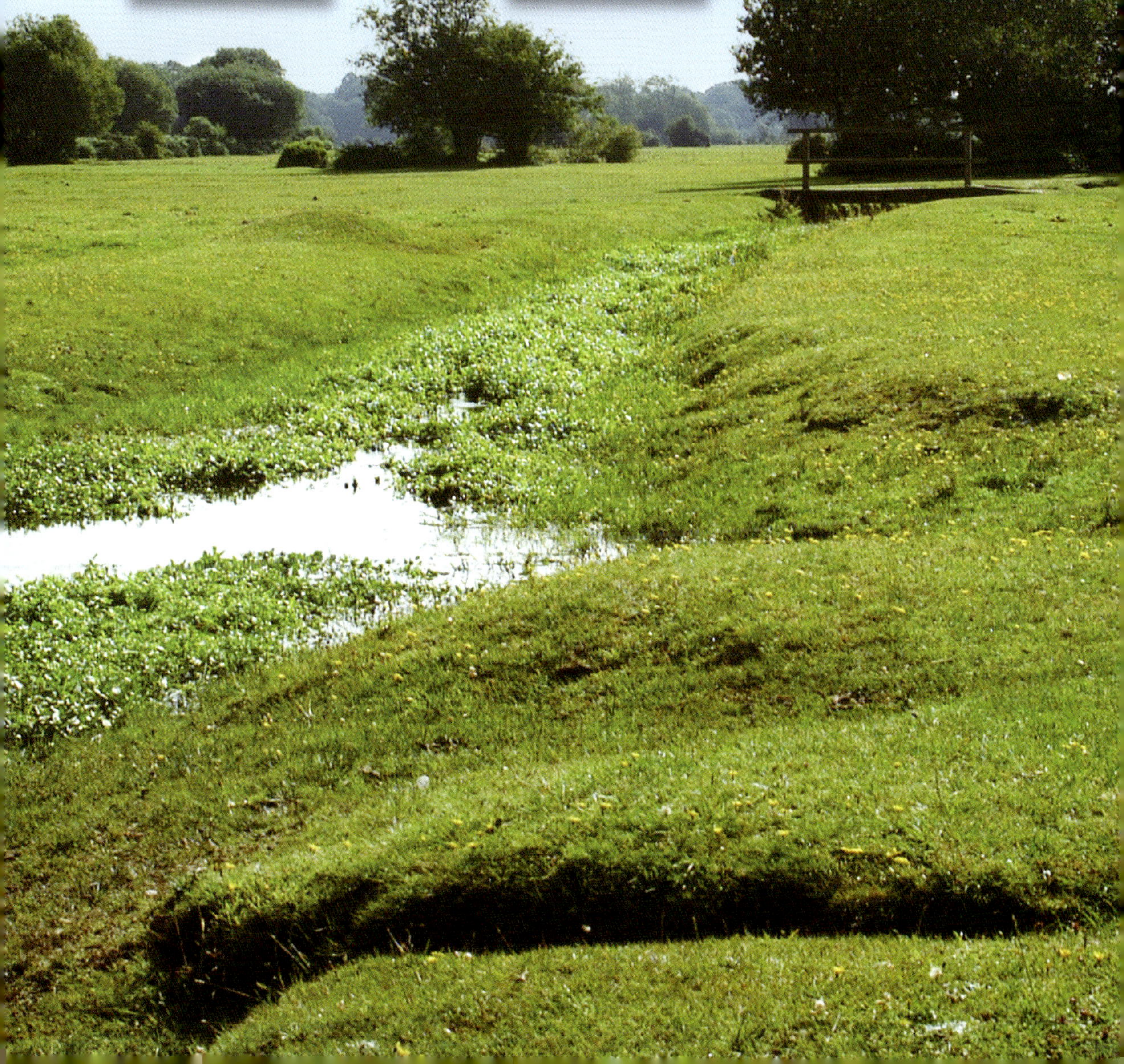

Heath

The large expanses of heathland appear quite natural, but they were first formed by Bronze Age people clearing the woodland that once covered the whole area of the New Forest. The fact that these areas have not been put to agricultural use is because the soil is generally sandy and infertile, so not suitable for growing crops. Nevertheless, it is widely colonised by typical heathland plants and is home to a variety of wildlife, much of which is not immediately visible. The low-growing vegetation makes good habitat for insects, and these in turn attract many different birds who feed on them. An assortment of grasses also provides seeds for birds.

In spring the gorse or 'furze' bushes burst into bloom with waves of bright yellow flowers. A summer walk on a forest track through the gorse is accompanied by the sharp staccato sounds of dried seed pods cracking open in the heat of the sun. Gorse bushes make good feeding sites for the Dartford warbler, which is rarely seen in other parts of Britain.

Come late summer, the rolling heathland turns purple as the heather blooms. Two kinds of heather grow on these tracts of land – the pale mauve ling and the brighter-coloured bell heather, with larger bell-shaped flowers. The heather provides an abundant supply of nectar for bees and other insects.

Interspersed with the heather in summer are expanses of green bracken, which in autumn die to a characteristic burnt orange colour. Nestling amongst the spreads of gorse, heather and bracken are small wild flowers such as tormentil and milkwort. Several species of wild orchids are found, and the rare wild gladiolus, which is a striking deep magenta colour, grows here too.

LEFT: *The rare Dartford warbler enjoys an ideal habitat in the New Forest, nesting in the long heather and feeding on insects.*

ABOVE: *There is rampant growth of gorse, which must be kept in check.*

RIGHT: *Heathland in summer; Bell heather* (left), *Ling* (right).

The heath is home to ground-nesting birds and it is important to keep to the established tracks to avoid disturbing them. The nightjar, for instance, is hard to see when camouflaged on its nest amongst the heather and bracken, though its distinctive churring sound can be heard after dark. In the daytime songbirds such as skylarks and the meadow pipit may often be heard. Towards the woodland edges there are tree pipits and, more rarely, woodlarks.

This is also hunting ground for raptors such as kestrel, buzzard and sparrowhawk. The hobby especially can be seen in these heathlands.

Managing the heathland

In winter and early spring, when the ground is damp, the Forestry Commission undertakes controlled burning of heathland areas. This prevents the growth of too much scrub vegetation and allows a new growth of grass, which is vital for the depasturing of animals. The dampness of the ground means that fallen seeds and the roots of plants are not affected. Burning heathland in this way results in the creation of diverse habitat structures essential for wildlife and which support many of the species for which the New Forest is deemed so special.

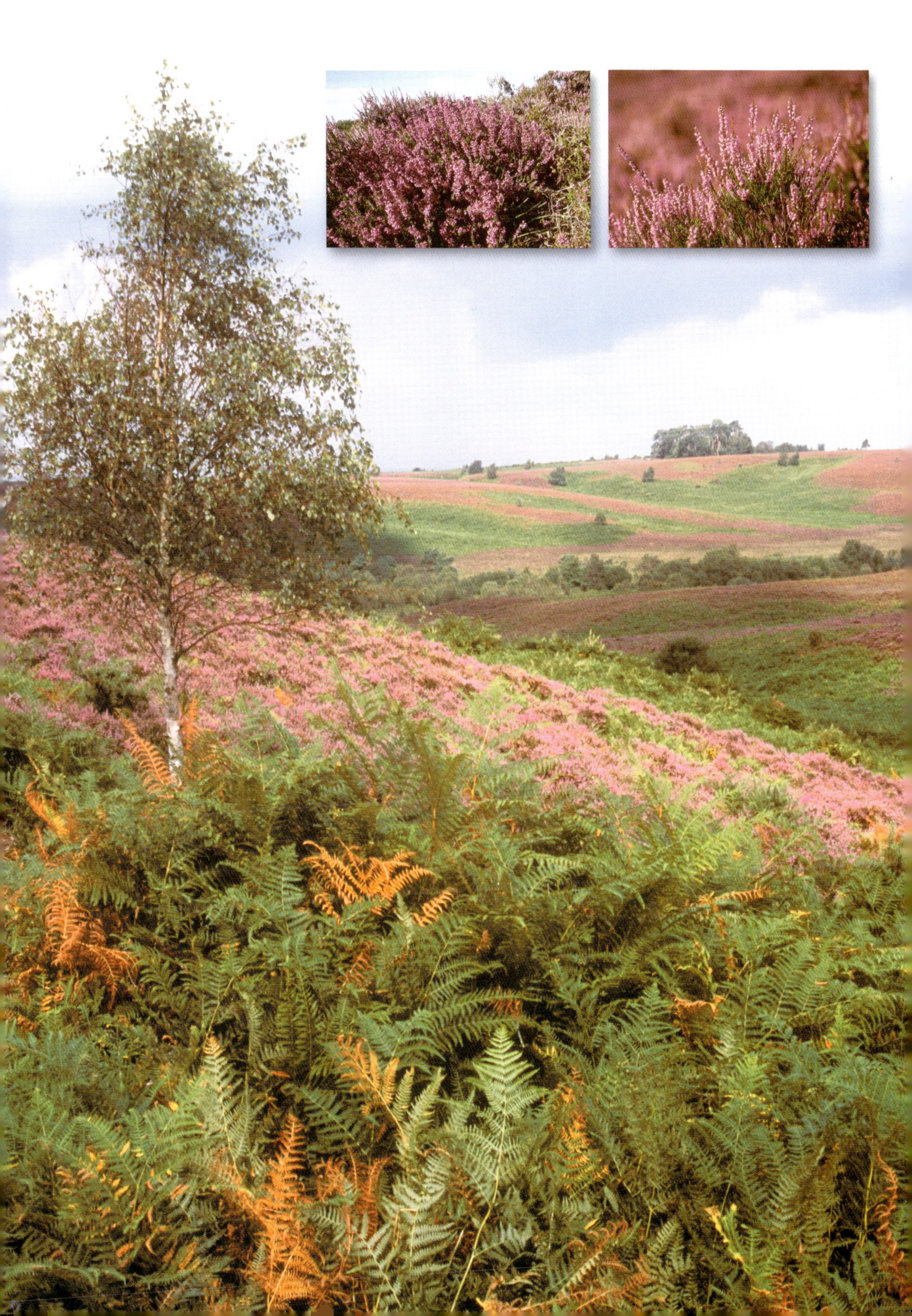

HEATHLAND REPTILES

Sandy heathland is an ideal location for reptiles and the New Forest has several species, one or two of which are quite rare and special to this area. All three native species of snake are found here – the adder, grass snake and smooth snake.

The adder or viper is Britain's only poisonous snake. It varies in colour, but may be identified by the dark zigzag pattern down the length of the grey-brown body and the 'V' marking at the back of the head. The pupil of the eye is slit shaped. The male grows to about 60 cm (2ft) and the female may be a little larger. An adder's diet consists of mice, voles, lizards, young birds and birds' eggs. In winter adders hibernate in holes under the sandy soil or a woodland bank, but they emerge in sunny weather. Snakes are cold-blooded animals, so like to stretch out and bask in the sun to regulate their body temperature. For this reason it is not a good idea to walk across heathland with open shoes or sandals.

The grass snake has a yellow ring underlined with black at the back of the head. The olive-brown body is marked with black bars along its length. The pupil of the eye is round. The female is larger than the male and usually measures about 100 cm (3 ft). Grass snakes are good swimmers and feed on frogs, toads, newts and fish, as well as young birds and eggs. In winter

ABOVE: *The venomous adder is recognised by the distinctive zigzag pattern along its back.*

BELOW: *The rare smooth snake.*

ABOVE: *The grass snake lays up to 40 eggs at a time.*

they hibernate under leaf litter or logs.

The smooth snake, so called because its scales are not ridged, is not often seen. Its main abode is under the vegetation on dry heathland and its grey-brown colouring acts as good camouflage. The body is marked with two rows of black bars or dots. This rare reptile is protected by law from any injury or disturbance. It eats lizards and small mammals, including shrews, which are distasteful to many other creatures. Most snakes lay eggs, but, like the adder, the smooth snake gives birth to live young.

Sometimes thought to be a snake, the slow-worm is actually a legless lizard. Fairly common in the forest, it is easily recognised by its grey-bronze body, with females a little lighter in colour than males. Slow-worms grow to about 50 cm (18 in) and eat slugs, snails and beetles. They hide in the woodland border, but also enjoy basking in the sun.

Two other lizards are found in the forest heathland – the common or heath lizard and the very rare sand lizard. The common lizard can be spotted darting away when disturbed. It is about 15 cm (6 in) long and a grey or brown colour with paler spots and an underside that is yellow or orange. It feeds on insects and spiders.

ABOVE: *Common or heath lizard.*

ABOVE RIGHT: *Sand lizard.*

The sand lizard is slightly larger, about 18cm (7 in). It has been reintroduced into the New Forest and is also found in the sandy Dorset heathlands. It is grey-brown in colour, with the male a little more greenish. Both sexes have dark spots with white centres along the back and sides. In the summer the sides and underneath of the male turn a bright green colour.

Brusher Mills

Harry Mills was one of the great characters of the New Forest. Born in 1840, he lived a solitary life in an old charcoal burner's hut and made his living from catching snakes. The nickname 'Brusher' is said to have come from his enthusiastic sweeping of the wicket between the innings at a local cricket pitch.

Brusher used simple means for catching a snake, pinning the creature down with a forked stick and then removing it to a tin with blunt tongs so as not to harm it. The tin was then safely placed in a sack. He sold the snakes to London Zoo, at one time receiving a shilling per snake. He caught both adders and grass snakes, but it was the grass snakes that were most useful since these were fed to other 'cannibal' snakes in the zoo.

Brusher died in 1905 and is buried in St Nicholas churchyard, Brockenhurst.

The New Forest Reptile Centre

This centre at Holidays Hill, near Lyndhurst, provides the opportunity to see and learn about native reptiles and amphibians and their conservation in the New Forest.

Mires and Bogs

Areas of bog in the forest are particularly important for much of this specialised habitat is disappearing in other parts of the country and also in the rest of Europe. These areas of mire occur in the low-lying valleys. Poorly drained, the result is a build-up of vegetation and other organic material that cannot decompose completely, so creating a peaty soil rich in nutrients.

Willow and alder grow in the boggier areas by streams, but as the land becomes more waterlogged there are fewer trees or shrubs. The valley mires are characterised by extensive growth of a number of plants. Here are found clumps of cross-leaved heather and purple moorgrass. Areas are colonised by tussocks of cotton grass, with stems that bear distinctive downy seed heads in summer. There are also bog myrtle bushes, which release their fragrance on warm summer days.

Cushions of sphagnum moss spread extensively, acting like sponges in holding moving water. In areas where there is less movement there is a rich abundance of different sedges. Among the flowers that grow in the boggy soil are the small pink-flowered bog pimpernel, the yellow bog asphodel and the bogbean. Rarer plants include the blue marsh gentian and the bog orchid.

LEFT: *Cotton grass thrives in acid bogland.*

BOTTOM LEFT: *Sundews derive nutrients from the insects that breed in these boggy areas.*

BELOW: *The forest ponds are frequented by ponies and cattle in summer when water is scarce.*

RIGHT: *Wetland and bog areas are the focus of much conservation. The New Forest is a prime location for observing dragonflies.*

BOTTOM RIGHT: *The bogbean flower grows above the water from a rhizome.*

BOTTOM FAR RIGHT: *The bog asphodel grows on peat.*

The pools formed within the bogland are habitat for many insects who live in the water or who breed there – this is a particularly important site for the southern damselfly. There are, too, countless smaller flying insects, which fall prey to the insectivorous plants that grow here. Butterworts exude a sticky substance on their leaves so that an insect alighting on a leaf is held fast. The leaf then curls around the insect and the plant absorbs the nitrogen from its body. Similarly, sundews capture insects on sticky pads on the ends of tiny hairs on their leaves. Once the insect is stuck the hairs close around it, tightly trapping the creature.

Bladderworts work in a different way. Their yellow flowers may be seen above the water, but most of the plant is below. Aquatic insects and organisms entering the tiny submerged bladders trigger an inner opening and are sucked into the plant.

The valley mires are used by birds, too, as they are relatively undisturbed here. Among the species that may be seen are lapwing, curlew and redshank.

The New Forest Show

Originally held as a one-day gathering for commoners to display livestock, the New Forest and Hampshire County Show is today a major agricultural and equestrian event that attracts thousands of visitors from all over the south of England, and competitors from even further afield. The show now lasts for three days and is held every summer at New Park, Brockenhurst, during the last week in July.

A wide array of attractions is on offer, but at its heart the show reflects and encourages rural life and country traditions that go back centuries.

Show rings feature horse and pony events, from show jumping, coach and carriage driving to heavy horse competitions and the judging of pony breeds. Parades and visiting team displays are also held here, interspersed with traditional championship competitions such as pole climbing.

Skirting the main rings is a plethora of agricultural tents for livestock and small animals, produce and foodstuffs, flowers and crafts. Trade stands cater for every aspect of outdoor life and country sports, and a wide choice of refreshments is available.

Areas are dedicated to traditional skills such as blacksmithing. Agisters and foresters are on hand to chat about their duties and discuss the ways of the forest with visitors and locals, and axemen demonstrate their woodland and forestry skills.

The exact nature of the show and its attractions varies slightly from year to year, but there is always plenty to interest anyone who has a general interest in country life and the New Forest in particular.

LEFT TOP: *Heavy horses, turned out for display work, are popular with the crowd.*

LEFT: *Rare breeds of livestock are welcomed at this agricultural show.*

FAR LEFT: *Beeswax is judged for its quality and colour.*

TOP LEFT and RIGHT: *Demonstrations of blacksmithing and woodland crafts.*

ABOVE LEFT and RIGHT: *Judging of the New Forest stallion class and champion livestock.*

RIGHT: *Skilled character craftsmen are always an attraction.*

Villages and Towns

Lyndhurst

Lyndhurst is the 'capital' of the New Forest and the Court of Verderers has always been held here, at The Queen's House. Today Lyndhurst is also the adminstrative centre of the New Forest District Council and home of the New Forest Museum and Visitor Centre.

The Queen's House is sited on the original manor house for Lyndhurst and was the official residence of the Lord Warden of the Forest. There are records of repairs to the building in the fourteenth century, but it was rebuilt in the seventeenth century under Charles II and substantially remodelled in the mid nineteenth century. Its name accords with the current monarch. The Verderers' Hall, which adjoins The Queen's House and in which the Court of Verderers sits, retains much of its Tudor appearance.

Lyndhurst's red-brick Victorian church of St Michael and All Angels is noteworthy for a number of features. It is richly decorated inside, with a large painted fresco of 'The Parable of the Wise and Foolish Virgins' by Frederic, Lord Leighton (1830–96) behind the altar. Among the noteworthy stained-glass windows are those designed principally by the pre-Raphaelite artist Edward Burne-Jones (1833–98), with Philip Webb, Ford Madox Brown and Dante Gabriel Rossetti, and made by the firm of William Morris.

In the churchyard is the grave of Alice Hargreaves (d. 1934). As a child, as Alice Liddell, she was the inspiration for Lewis Carroll's 'Alice in Wonderland'.

Just outside the village is Swan Green. In the summer cricket is played on this part of the forest in an idyllic setting enhanced by a row of picturesque thatched cottages.

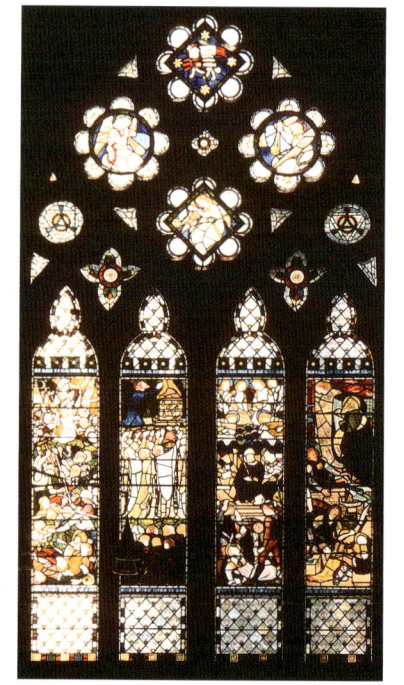

LEFT: *Pre-Raphaelite stained-glass window, St Michael's, Lyndhurst.*

BOTTOM LEFT: *Leighton fresco in St Michael's, Lyndhurst.*

TOP LEFT: *Bolton's Bench, Lyndhurst.*

TOP RIGHT: *Cottages at Swan Green, Lyndhurst.*

BELOW: *Gravestone for 'Alice', St Michael's.*

Brockenhurst

The attractive village of Brockenhurst has long been a settlement and the Domesday Book makes mention of four small manors in the area. Its location in the heart of the forest is enhanced by the various streams or 'waters' that flow through and around it, and there is a picturesque 'watersplash' at the end of the village. The land-owning Morant family lived at Brockenhurst Park.

Brockenhurst is a thriving centre, helped by the building of the railway and station in 1847.

St Nicholas Church is probably the oldest church in the forest and may be considered mainly Norman, though it was most likely built on an early Saxon site and there are even some Roman materials. It is a treasure trove of medieval architectural features, with a font dating to the twelfth century, and with many additions from later centuries. 'Brusher' Mills, the snakecatcher, is buried in the graveyard.

Balmer Lawn, with Lymington River flowing at its side, is popular with families for recreation. The hotel formed part of a hospital during World War I, first for Indian soldiers shipped back to England and then for wounded New Zealand troops.

Rhinefield House, now a hotel, was rebuilt in the late nineteenth century by the Walker-Munro family. Its setting gives far-reaching views across open forest on one side, while on the other it is reached by the Ornamental Drive, an impressive avenue of rhododendrons and specimen fir and redwood trees.

The New Forest and Hampshire County Show is held annually at New Park, Brockenhurst.

TOP LEFT: *Watersplash, Brockenhurst, in winter.*

TOP RIGHT: *St Nicholas church, Brockenhurst.*

RIGHT: *These redwoods in the Ornamental Drive are the tallest trees in the forest.*

BELOW: *Rhinefield House.*

Beaulieu

This bustling little village, situated on the Beaulieu River, is part of the Beaulieu estate. The history of Beaulieu (meaning 'beautiful place') dates to before 1204, when King John granted land to Cistercian monks to build an abbey on the site of a royal hunting lodge. A thriving community grew around the magnificent religious buildings and Beaulieu became known for its scholarship and expertise in agriculture. With the dissolution of the monasteries by Henry VIII, however, the religious community was forced to leave. The estate was bought in 1538 by Sir Thomas Wriothesley, later to become 1st Earl of Southampton. His descendants, the Montagu family, own Beaulieu today.

Many of the monastic buildings were demolished, but the Great Gatehouse of the Abbey became Palace House. Although a family home, parts of the house and gardens are open to the public. Other surviving buildings that can be visited are the Domus or lay brothers' dormitory, which now houses an exhibition illustrating early monastic life, and the ruins of the Abbey Cloisters. The Refectory became the parish church and is regularly used for worship.

Beaulieu is also home to the National Motor Museum. This collection of some 250 vehicles presents a history of motoring from its earliest days to the present, with examples of cars from every era of driving. It includes famous land speed record-breakers such as *Bluebird* and *Golden Arrow*, and many examples of legendary classic and racing cars. Dedicated car rallies and memorabilia sales are held throughout the year.

LEFT: *Beaulieu village, part of the Beaulieu estate.*

BELOW: *Beaulieu village from across the mill pond.*

RIGHT: *The National Motor Museum is a magnet for motoring enthusiasts.*

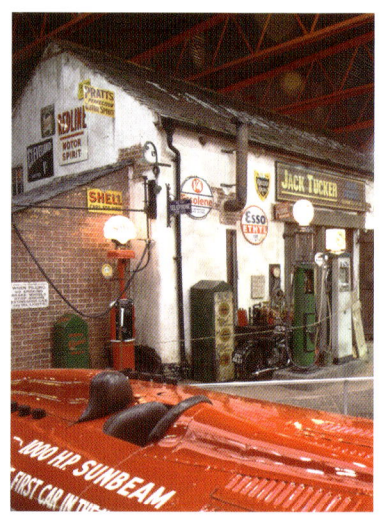

RIGHT: *Buckler's Hard viewed from across the Beaulieu River.*

BELOW: *Exbury Gardens are world famous for the Rothschild collection of rhododendrons and azaleas.*

Buckler's Hard

Just over two miles down the Beaulieu River from Beaulieu is the maritime museum village of Buckler's Hard. This eighteenth-century village consists of two facing rows of terraced cottages on a grassy slope that leads down to the river. Buckler's Hard is part of the Beaulieu estate and was built by the 2nd Duke of Montagu, who originally planned it as a free port.

From the mid eighteenth century this was the scene of important shipbuilding activity. Many of the Royal Navy's great wooden ships were built here, including Nelson's ship *Agamemnon*, *Swiftsure* and *Euryalus*, all three of which took part in the Battle of Trafalgar in 1805. Today the house of the Master Builder, Henry Adams, is a hotel, and visitors may see a reconstruction of his office. Other reconstructions show cottage life at the time.

Buckler's Hard was also important during World War II. Parts of the Mulberry Harbours were constructed here prior to D-Day, and naval and military personnel gathered along the river prior to the Normandy landings in 1944.

Exbury

On the other side of the Beaulieu River from Beaulieu is the Exbury estate. In 1919 Lionel de Rothschild bought Exbury House with the specific purpose of establishing a woodland garden. His special interest was in rhododendrons, which flourish on the area's acidic soil. He laid an extensive series of paths and ponds and an underground system of pipes to ensure efficient watering throughout the garden. Lionel died in 1942, and since World War II his son Edmund has developed Exbury Gardens and its collection of rhododendrons, azaleas and camellias to world-class fame. Many new varieties have been cultivated here and Exbury plants are exported all over the world.

During World War II Exbury House was requisitioned by the Royal Navy as HMS *Mastodon* and played a central role in the preparation of crews for the many types of landing craft used for the Normandy landings.

LEFT: *'Mare and Foal'* by Priscilla Hann, The Furlong Centre, Ringwood.

RIGHT: *Cricket, Godshill.*

Ringwood

The small town of Ringwood on the River Avon lies just outside the forest perambulation on the western side of the forest. Its name in the Domesday Book was 'Rincvede', but it had become Ringwood by the thirteenth century, perhaps from an earlier Saxon name. The town's location between river and forest gave rise to its development as a market town and the weekly market dates back to 1226 when a charter was granted to the Lord of the Manor of Ringwood.

In 1685 the Duke of Monmouth, illegitimate son of Charles II, was captured near here after his defeat at the Battle of Sedgemoor, and was held in Ringwood prior to being taken to London, where he was executed.

The Morant family of Brockenhurst bought the manor of Ringwood in the eighteenth century and established a cottage industry that produced knitted stockings and gloves.

The River Avon is famous for its trout fishing, but the river also provided water for making beer and the town once boasted several breweries. Brewing has been successfully revived in Ringwood.

The Meeting House, a Unitarian chapel built in 1729, is a simple, but imposing, red-brick building and now houses a small museum.

ABOVE: *The Meeting House, Ringwood.*

LEFT: *St Peter & St Paul and the modern market.*

BELOW: *The medieval bridge, Fordingbridge.*

Fordingbridge

As its name suggests, the old market town of Fordingbridge is situated beside the river Avon, just outside the north-western edge of the New Forest. There are, however, holdings in the area with rights of common. Several forest waters feed into the Avon, which is renowned for its coarse fishing.

Fordingbridge was a Saxon settlement, although Iron Age coins have been found at Godshill to the east of the town. St Mary's church has Saxon origins, but its earliest surviving details date from the twelfth century and the building has been reconstructed with many changes and additions in the succeeding centuries.

Much of Fordingbridge was destroyed by fire in 1702, with the loss of the medieval and half-timbered Tudor architecture that made up the heart of the town. Only the church and the prominent arched medieval bridge survive from the early period.

The artist Augustus John (1878–1961) lived in this area for many years.

The New Forest

Burley

This popular village attracts many tourists, drawn by a plethora of gift shops and tea rooms. The area has long been inhabited and there are far-reaching views from Castle Hill, the site of an Iron Age hillfort.

Burley was an Anglo-Saxon settlement, though first recorded in the twelfth century with mention of Peter de Burley in the reign of King Henry II. The current Burley Manor was rebuilt in the mid nineteenth century and is now a hotel.

The Queen's Head public house dates to the seventeenth century and was an important gathering place for information when smuggling was an integral part of the life of the village.

Legend and witchcraft have played a part in folklore here, too. Burley Beacon is reputed to have been the lair of the legendary Bisterne Dragon, which demanded a pail of milk each day and was eventually slain by a local knight. More recently, in the 1950s and early 60s, the village was known as the home of Sybil Leek, a 'white' witch, who named the gift shop A Coven of Witches.

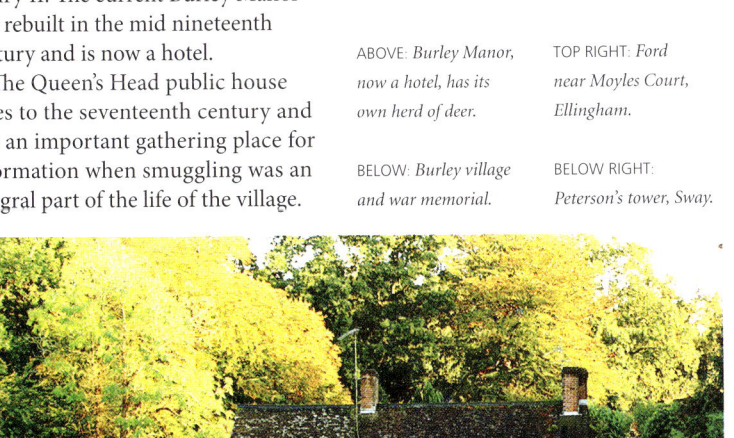

ABOVE: *Burley Manor, now a hotel, has its own herd of deer.*

BELOW: *Burley village and war memorial.*

TOP RIGHT: *Ford near Moyles Court, Ellingham.*

BELOW RIGHT: *Peterson's tower, Sway.*

Ellingham

The Avon Valley path enters the New Forest for a short distance around the area of Ellingham, where Dockens Water runs towards the river.

Moyles Court, now a school, was the home of the redoubtable Dame Alice Lisle. Accused of harbouring supporters of the Monmouth rebellion in 1685, she was sentenced to death by Judge Jeffreys at the Bloody Assizes in Winchester. Although 70 years of age, she was publicly beheaded.

Sway

Sway's most obvious feature is its landmark tower, which can be seen for miles across the tree tops. In fact, close up, there is a smaller, earlier tower, overshadowed by the taller one of 220 ft (67 m). Both were constructed in the 1880s by retired Judge Peterson as an experiment to prove the reliability of shuttered concrete.

In 1847 Captain Marryat used Sway as the backdrop to his story *The Children of the New Forest*.

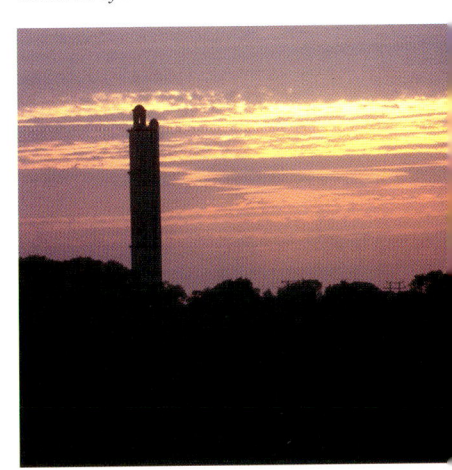

Minstead

The village of Minstead grew up around a manorial estate that pre-dated the Norman Conquest. Legend has it that William Rufus spent the night before his death in this area of the forest, near the site of Iron Age earthworks. The Rufus Stone, which commemorates the king's demise, is at nearby Canterton.

All Saints at Minstead is one of the most interesting churches in the forest. Parts of the building date from the thirteenth century, but the site is believed to have been used as a place of worship long before this. Much has been added over the centuries and the church offers a variety of architectural styles. A gallery for musicians and choir dates from the late eighteenth century, and the Victorians added another for the 'poor'. In contrast to the latter, there are two private 'parlour' pews, one of which still has its fireplace. These were built in the nineteenth century for the local squire and his family.

The churchyard holds the grave of Sir Arthur Conan Doyle (1859–1930), who knew the area well. His novel *The White Company* is set around the New Forest, with mention of Minstead.

LEFT: *All Saints church, Minstead.*

BOTTOM LEFT: *Saxon stone font and seventeenth-century three-decker pulpit in Minstead parish church.*

BELOW LEFT: *The Trusty Servant, Minstead.*

Situated next to the village green is the village pub The Trusty Servant. Its unusual inn sign is a copy of a well-known medieval painting held in Winchester College.

RIGHT and BELOW: *The Royal Oak, Fritham.*

Fritham

Fritham, with its agricultural settlement, may be mentioned in the Domesday Book as 'Ivare'. Today this hamlet towards the north of the forest retains its traditional rural character. However, the quiet setting belies the fact that from 1863 to 1921 the area was the location of a thriving gunpowder factory that employed a large workforce. Locally produced charcoal was used as an ingredient of gunpowder and water was provided by a chalybeate (iron-rich) spring. The source was enlarged and still remains as Irons Well.

Generations of commoners, foresters and visitors have gathered to sup and discuss matters at The Royal Oak, a small thatched 'parlour' pub that has altered little over the years.

The plain between Fritham and Stoney Cross was used as an airfield during World War II.

Lymington

Lymington lies outside the boundary of the New Forest, but it is closely allied with the life of the forest (ponies used to visit its High Street). Its location on the The Solent ensures its popularity as a maritime and leisure centre, with a Saturday market an added attraction for many.

Recorded as 'Lentune' in the Domesday Book, Lymington was a small port when the Normans arrived, but its prosperity grew with the export of salt extracted from the surrounding salt marshes. Shipbuilding was developed in the Middle Ages and today boat building continues to be of commercial importance. New marinas cater for a large yachting and sailing fraternity.

Much of the architecture of the town is Georgian and Victorian. The cobbled Quay Hill, with its bow-windowed buildings, leads from the High Street to the quay by the Lymington river. Quaint town houses and cottages retain their original features and make a perfect backdrop to tales of smuggling in days gone by. The press-gangs were active here, too.

A footpath leads along the 'sea wall' to Keyhaven.

BELOW: *The quay at Lymington, a popular sailing centre.*

ABOVE: *Cupola, St Thomas's, Lymington.*

RIGHT: *Town houses, Lymington.*

BELOW: *Quay Hill, Lymington.*

BELOW RIGHT (and INSET): *William Gilpin, vicar of Boldre, is remembered as teacher, reformer, artist and writer.*

Boldre

Just north of Lymington lies Boldre, a village of 'scattered' cottages with the church of St John the Baptist on a hill at its heart. William Gilpin was vicar here from 1777 to 1804. His series of drawings and writings illustrate the tours he made around rural Britain and 'Remarks on Forest Scenery' describe the New Forest. Gilpin is buried in the churchyard.

There is a memorial in the church to HMS *Hood*, sunk in 1941, with the loss of over 1,400 lives.

Keyhaven and Hurst Castle

Adjoining the sea wall of The Solent, the salt marshes at Keyhaven provide breeding grounds for a variety of birds, and in the winter this coastal reserve also attracts large flocks of wildfowl and wading birds. Mudflats and water channels interspersed between the marshy vegetation are alive with tiny organisms that are an important food source for waders and other wildfowl. Among the bird species that may be seen are black-headed gulls, terns, oyster catchers, redshanks, dunlin and lapwing.

Several types of plants thrive, including cord grass, glasswort, sea purslane and sea aster, and the rarer golden samphire.

The salt marshes are protected by the shingle of Hurst spit that extends for 1½ miles. At the end of the spit is Hurst Castle, which may also be reached by ferry from Keyhaven. Dating from 1544, this rather forbidding building formed part of Henry VIII's series of fortresses built for coastal defence, as was Calshot Castle at the mouth of Southampton Water. In 1648 King Charles I was imprisoned here, en route from incarceration at Carisbrooke Castle on the Isle of Wight, before being taken to trial and execution in London.

Hurst Castle was updated during the Napoleonic wars, and was substantially enlarged later in the nineteenth century in response to the development of iron-hulled ships. It was also manned during both World Wars and it formed part of the coastal artillery defences until 1956.

The lighthouse, which functions automatically, enables shipping to navigate the channel between the Shingles Bank and the Needles.

ABOVE: *View of Hurst Castle across the mudflats at Keyhaven. The Isle of Wight can be seen in the distance.*

RIGHT: *Dunlin, in their winter plumage, feeding in the early morning on the Keyhaven mudflats.*